# INTRODUCTION

These guidelines are intended for the design of cable-stayed bridges.

They are intended as performance specifications to be used with any national code.

However, national codes for live loads and other important provisions differ. Yet safety level depends on all loads and resistance parameters. Therefore, a more complex treatment of the design requirements may be required under certain conditions.

The guidelines in these papers are the best estimates by the ASCE Committee on Cable Suspended Bridges. They have been developed based on current practice and the latest thinking for this type of bridges. Equations given in SI-UNITS with equations in parenthesis in English Units.

# CONTENTS

| | Page No. |
|---|---|
| CHAPTER 1 - Safety Factors, Loads, and Impact | 1 |
| CHAPTER 2 - Cables and Anchorages | 11 |
| CHAPTER 3 - Stability of Cable-Stayed Bridges | 21 |
| CHAPTER 4 - Dynamic Wind Loads | 26 |
| CHAPTER 5 - Special Considerations for Railroads | 41 |
| CHAPTER 6 - Seismic Design | 45 |

# CHAPTER 1

## SAFETY FACTORS, LOADS, AND IMPACT

### 1.0 GENERAL

The loads given in this chapter are intended for general guidance only (including wind, ice, stream flow, etc.). Maximum deflections and rotations may likewise vary with national codes. In all cases, loads must be considered in conjunction with allowable stress and/or load factors.

### 1.1 DEAD LOADS

The dead load shall consist of the weight of the entire structure, including the roadway, sidewalks, car tracks, pipes, conduits, cables, and other public utilities.

For roadway bridges it shall include parapets, medians, separate and future wearing surfaces, and railings.

For railroad bridges it shall include tracks, ties, ballast, deck and walkways, and railings.

For stay forces, the weight of steel tendon, sheathing and grout, when present, shall be included as dead load in the stay or cable.

### 1.2 LIVE LOADS

It is recommended that live load follows the loads given in the governing code for that structure.

Number of traffic lanes, railroad tracks, impact, longitudinal forces, centrifugal forces, reduction in load intensity for multiple lane or tracks, railing loading, sidewalks loading, etc., according to codes. In addition to these loads, checkerboard loading to produce maximum torque should be investigated. Impact shall be computed in accordance with the national code used. Members to be included are superstructure members and stays. Piers and towers are excluded from impact.

### 1.3 WIND LOADS

All wind loads in this chapter are "static," for "dynamic" wind loads, see Chapter 4. The wind loads shall be as defined in the national code, however, the intensity applied horizontally at right angles to the longitudinal axis of the structure shall at least be:

$$W_h = \frac{Z^{0.2} \times V_{10}^2 \times C_h}{25.6} \quad Pa$$

$$(W_h = \frac{Z^{0.2} \times V_{30}^2 \times C_h}{600} \quad lbs/ft^2)$$

Where   $Z$ = Height in m (feet) of the top of the floor system above ground level, or 10m (30 feet), whichever is greater.

$V_{10}$ ($V_{30}$) = Fastest mile wind speed, in km/hr (mph), at 10m (30 feet) above ground, for 100-year mean recurrence interval (see Figure 1 for use in the U.S.A.). For other countries reference is made to similar statistical information.

$C_h$ = Shape factor for horizontal wind load. $C_h$, unless wind tunnel data confirms different values, shall be at least:

for plate and box girders $C_h$ = 1.5
for trusses (applied to   $C_h$ = 2.3
one plane only)

See: "Recommended Design Loads for Bridges," ST 7, July 1981, ASCE Journal of the Structural Division.

For $Z$ = 10m (30 feet), $V_{10}$ = 161km/hr ($V_{30}$ = 100 mph) and values given for $C_h$, the wind load is 2,400 Pa (50 lbs/ft$^2$) and 3,600 Pa (75 lbs/ft$^2$) respectively. For wind loads under different skew angles, the lateral and longitudinal load components may be found by force components or using the same ratios as given in AASHTO (American Association of State Highway and Transportation Officials. Standard Specifications for Highway Bridges 14th Ed. 1989).

The wind loads on cables may be found by the same equation given above using $C_h$ = .7. Proper damping of cables must be provided, due both to vortex shedding and to oscillations due to variations in cable tension.

## 1.4 WIND ON LIVE LOAD

This load shall be applied 1.8m (6 feet) above the deck for highway bridges. For railroad bridges see Chapter 5. The magnitude of this load is:

$$WL = \frac{Z^{0.2} \times V_{10}^2 \times C_{h1}}{25.6} \quad Pa$$

$$(WL = \frac{Z^{0.2} \times V_{30}^2 \times C_{h1}}{600} \quad lbs/ft^2)$$

Where $C_{h1} = 1.0$

and $V_{10}$ ($V_{30}$) = shall not exceed 89 km/hr (55 mph).

The magnitude of the exposed area shall be the loaded length of superstructure multiplied by a height of 3m (10 feet).

## 1.5 WIND ON SUBSTRUCTURE

The horizontal wind loads per square foot of exposed area of substructure is calculated from the expression under Section 1.3 with:

$C_h = .7$ for circular
$C_h = 1.4$ for octagonal
$C_h = 2.0$ for rectangular sections

For wind directions assumed skewed to the substructure, this force shall be resolved into components or may be applied according to AASHTO.

## 1.6 VERTICAL WIND - OVERTURNING FORCES

An upward force shall be applied at the windward quarter point. This force per unit area of deck is calculated for the expression in 1.3 where $C_v$ is substituted for $C_h$. $C_v$ may be assumed to be $C_v = .8$.

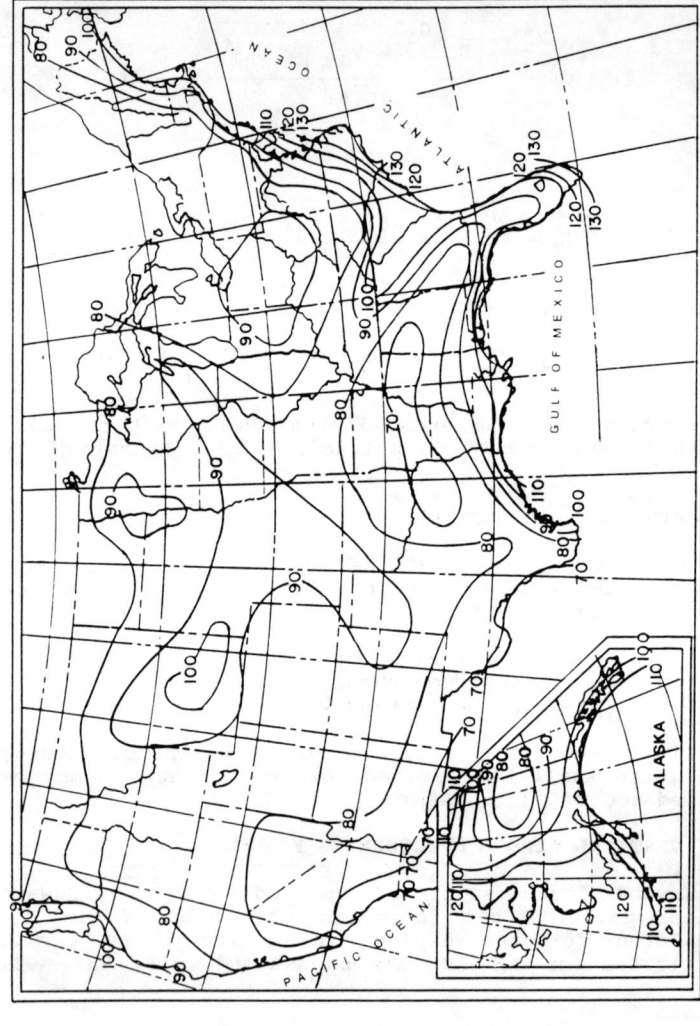

FIG. 1 - ANNUAL EXTREME FASTEST-MILE SPEED 30 FT ABOVE GROUND, 100-YR MEAN RECURRENCE INTERVAL. EXTRACTED FROM H.C.S. THOM: "NEW DISTRIBUTIONS OF EXTREME WINDS IN THE UNITED STATES" JULY 68, ASCE JOURNAL OF THE STRUCTURAL DIVISION.

| Cross-section | Strouhal No. S |
|---|---|
| 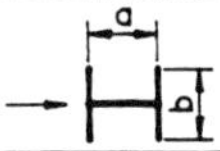 | 0.12 for $\frac{a}{b} = 1.0$ |
| 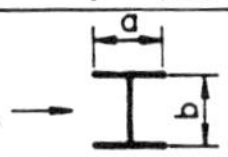 | 0.14 for $0.25 < \frac{a}{b} \leq 1.0$ |
| 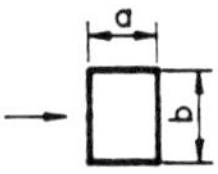 | 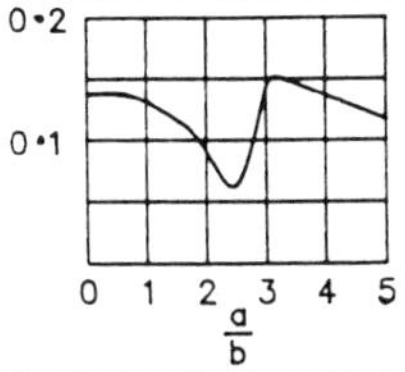 |
|  | 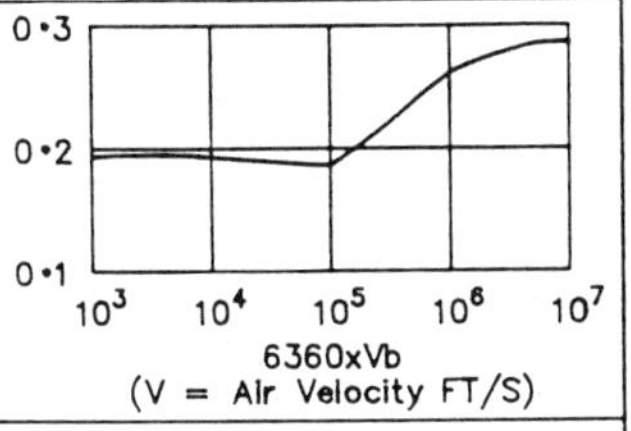 |
| 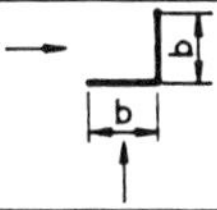 | 0•14 |
| 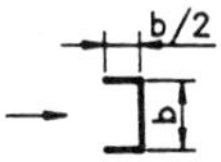 | 0•14 |
| 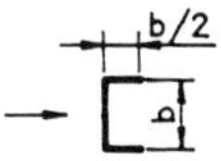 | 0•15 |

Table 1 — Strouhal Number for Various Cross Sections

Lower values of $C_v$ and different points of action may be used if justified by wind tunnel test.

## 1.7 ACROSS WIND VIBRATION OF STRUCTURAL MEMBERS

Slender structural members such as hangers, truss members, spandrel columns, and cables, shall be designed to resist resonant vibrations caused by vortex shedding. The critical wind velocity is determined by:

$$V_{crit} = 3.6 \frac{fb}{S} \text{ km/hr}$$

$$(V_{crit} = 0.68 \frac{fb}{S} \text{ mph})$$

S = Strouhal number (as given in Table 1)
f = frequency in Hertz
b = characteristic width of cross section in m (feet)

$V_{crit}$ shall not be less than $0.91 \times Z^{0.1} \times V_{10}$ ($0.81 \times Z^{0.1} \times V_{30}$). Consideration shall be given to the effect of wind loads on cables and to the need for mechanisms to dampen cable oscillations. Cable oscillations also occur due to changes in cable tension caused by live load. During rain, water rivulets running down the cable may, in combination with wind, cause disturbing cable oscillations. Additional wind considerations are given in Chapter 4 - Dynamic Wind Loads.

## 1.8 THERMAL FORCES

In addition to the uniform thermal forces normally used for structures in a particular climatic area, temperature differential between cables and deck or cables and pylons must be considered. A minimum of 10°C (18°F) temperature differential should be applied for white taped or painted stays. Site specific numbers for temperature differentials for other colors should be developed and may be as high as 20°C (36°F) for black stays. Thermal gradients for box girders must be considered. A ±10°C (±18°F), gradient through deck is recommended. A ±8°C (±15°F) differential between the inside and outside of the box should be investigated, but not simultaneously with the deck gradient. For open deck structures, a ±5°C (±9°F) gradient through the deck should be considered. Furthermore, reference is made to "National Cooperative

Highway Research Program Report 276 - Thermal Effects in Concrete Bridge Superstructures."

## 1.9 SNOW AND ICE ON SUPERSTRUCTURES

In areas where it can be expected that the bridge may be closed to traffic due to accumulated snow, the bridge should be designed for such loading conditions.

## 1.10 FORCES FROM STREAM CURRENT AND FLOATING ICE

Forces from stream flow may be assessed by the equation:

$$P = 515 \times K \times V^2$$

$$(P = K \times V^2)$$

P = pressure in Pa (pounds per square foot)
V = water velocity m/sec (ft/sec)
K = form factor; 1.4 for square ends
                 .7 for circular piers
                 .5 for angle ends 30° or less

When currents may become of such magnitude that scour is possible, proper scour protection is required. Ice forces on piers shall be selected with regard to site condition and mode of ice action such as dynamic or static ice pressure, ice jams, or static uplift. Dynamic ice forces may be calculated by:

$$F = C_n \times p \times t \times w$$

F = ice pressure in N (pounds)
$C_n$ = nose coefficient (1.0 = vertical nose .5 = 30° to 45° nose inclination to vertical)
p = ice strength .7 to 2.75 N/mm$^2$ (100 to 400 pounds per square inch)
t = ice thickness mm (inch)
w = width of pier in contact with ice mm (inch)

## 1.11 FORCES FROM SHIP COLLISION

Piers on navigational waters should be designed for ship impact unless otherwise protected by fendering systems, etc.

## 1.12 BUOYANCY AND EARTH PRESSURE

Buoyancy and earth pressure shall be considered where it affects the design of substructure or superstructure.

## 1.13 EARTHQUAKES

In regions where earthquakes should be anticipated, structures shall be designed to resist earthquake motions considering the relationship of the site to active faults, seismic response of soils, and the length of the structure. It may be done by an equivalent static force method, response spectrum method, or multi-mode spectral procedures. For complex structures a time-history response analysis may be considered.

Multimode spectral procedures should be considered especially if the acceleration coefficient for the bridge site is greater than 0.2. Unlike classical suspension bridges, vibrations of cable-stayed bridges cannot be categorized as solely vertical, lateral or torsional; instead, a three-dimensional motion is associated with almost every mode of vibration. For long structures, it is furthermore obvious that the structure is subjected to different motions at each of its foundations. Additional consideration is given in Chapter 6 - Seismic Design of Cable-Stayed Bridges.

## 1.14 CONSTRUCTION LOADS

Loads applied during construction must be investigated. Some recent construction methods like balanced cantilever construction leave the structure more vulnerable during construction stages than in its finished stage. Wind oscillations about a horizontal and/or vertical axis at the tower must be investigated. For structures which are constructed or designed for a balanced loading condition due to dead load, an unbalanced load, up on one arm and down on the other must be added. The magnitude of the unbalanced load shall be stipulated on basis of construction tolerances and variation of material densities. Superstructure towers and/or piers should be designed for this unbalanced load which is a part of the dead load. In areas with earthquake, the risk should be evaluated. The maximum unbalanced moment due to sequential addition of segments or panels is the moment produced by one segment or panel, plus erection equipment needed for construction (traveler, forms, cranes, etc.). In addition to these loads, an additional 480 Pa (10 psf) shall be applied on the loaded cantilever and 240 Pa (5 psf) on the second cantilever to account for unanticipated construction equipment.

During construction it is customary to apply 100% wind on one cantilever and 50% on the other for design purposes. Due to infrequent occurrence, seismic loadings are usually strongly reduced for the construction stage (see Chapter 6).

## 1.15 LOADING COMBINATIONS

Since live loads are following national codes it is only natural that loading combinations, allowable stresses and/or load factors follow the same code for consistency.

## 1.16 FATIGUE LOADS

Again it is recommended to follow the national codes for live load to be extended for fatigue loads.

Fatigue loads for cables are given in "Recommendations for Stay-Cable Design and Testing" by the PTI Ad Hoc Committee on Cable-Stayed Bridges.

Cable fatigue stresses are discussed in Chapter 2 - Cables.

## 1.17 MAXIMUM DEFLECTION AND ROTATION

There is no direct set limitation for deflections, however, it is part of the P-$\delta$ analysis. Maximum slope changes due to live load are more of a service consideration and dictated as follows for the use of bridges:

  Maximum slope change for truck loading:  2.0% (1.15°)

For railroad see Chapter 5.

## 1.18 STRUCTURAL STABILITY AND LINEARITY

Cable structures are ones in which loads are transmitted from the deck to the towers by tensile forces with no compression allowed. Changes in loads cause changes in cable geometry. The literature contains ample guidance with respect to suspension bridges. For cable-stayed bridges the minor sag effects may be accommodated by the use of an equivalent modulus of elasticity. Because the deck structures often are slender (large span/depth ratio), second order effects caused by the deck thrust must be accounted for. The most common method is to use a set of uniform and concentrated loads for linear combination of maximum live load cases. The second order effects (P-delta) will be determined separately for maximum loading cases along the girder.

The tower structure should likewise be investigated for second order effects.

There are computer programs on the market which will give the second order effects, from which the $n^{th}$ order may be predicted. Non-linear programs are available, but may become rather tedious for a full investigation.

The deck structure must be properly investigated against buckling, and it should be noticed that proper interaction equations for bending, thrust, and shear must be used. Reference is made to "Proposed Design Specifications for Steel Box Girder Bridges" - Report No. FHWA-TS-80-205.

Further discussion is made in Chapter 3 - Stability of Cable-Stayed Bridges.

# CHAPTER 2

# CABLES AND ANCHORAGES

## 2.0 GENERAL

The following types of cables are generally used in cable stayed bridges:

1. Parallel bars
2. Parallel seven wire strands
3. Parallel or semi-parallel wire cables
4. Locked coil strands and ropes
5. Helical galvanized bridge strands

The first two types are extensively used in prestressed concrete structures and when applied as cables in stayed girders bridges, a very similar installation technology is used. The cable is installed in a polyethylene tube or steel pipe which is then filled with grout. The grout adds about 30% to the weight of the cable. This adds to the cable sag and decreases its stiffness.

The three latter types have been used for a long time in structures supported primarily by tension members. The guiding principle here has always been ease of inspection and maintenance.

## 2.1 CABLE STRENGTH

Under normal design load (dead plus live plus impact) the static stress in the cable shall not exceed 45% of the guaranteed ultimate tensile strength (GUTS) of the cable.

The design fatigue stress range of the cable is determined by tests. The fatigue strength of the cable is assumed to be the stress range at which 5% of the wires have failed when the cable is subject to an average tensile strength of 45% GUTS. The number of cycles used in the test is normally $2 \times 10^6$ for highway bridges and $1 \times 10^7$ for railway bridges.

The design fatigue stress range is defined as:

$$\Delta\sigma_{Fat} = \frac{\Delta\sigma_F}{\gamma_{Fat}}$$

Where $\Delta\sigma_{Fat} = \sigma_{max} - \sigma_{min}$

$\Delta\sigma_F$ = fatigue stress range as determined by tests

$\gamma_{Fat}$ = factor of safety may vary from 1.25 to 1.50

Normally two curves are used to determine the allowable fatigue stress range.

1. The Wohler Curve. The stress range is the ordinate, the number of cycles the abscissa. The tensile force in the cable does not enter the picture.

2. The Smith Diagram. The stress range is the ordinate, the average stress is the abscissa. The number of cycles are given.

Examples of these curves are shown in the appendix in Figures 1 through 5.

It is customary to specify both static and fatigue tests for the cables in the steel plant before accepting delivery. The easiest things to test are single wires and bars and from these results compute the cable strength taking into account the "group effect." This means that the combined strength of several tensile members is less than the sum of the individual members.

A cable made up of several seven wire strands have two "group effects." The strength of a strand is less than seven times the strength of one wire and the strength of the cable is less than the sum of the individual strands.

The manufacturer's catalog will usually suggest what the value of those "group effects" are for preliminary design purposes, but the final design strength is determined by tests.

Most engineers will normally specify that tests be made on full size cables with anchorages attached at each end. It is sufficient to use three cables for each type of test.

In fatigue tests the median cable stress shall be at least 45% GUTS. The cable must break outside the anchor socket.

The length of the test cable is usually about 4 meters (13 feet).

## 2.2 DESCRIPTION OF CABLES

E - Young's modulus in $N/mm^2$ (ksi = kips per square inch).

GUTS means guaranteed ultimate tensile strength.

Creep of steel can be neglected since the maximum stress under normal loading (dead plus live plus impact) must not exceed 45% GUTS.

Whenever an applicable ASTM specification applies, it is mentioned. Normally, it applies only to plain steel tension elements, which means they are not galvanized.

Since the strength of these elements is constantly subject to change in metallurgy and protective corrosion coating, testing is necessary and the standard specifications are of little value.

In what follows reference is made to some ASTM specifications and some typical design stresses that may be used in a preliminary design. But in the final design, actual values obtained from tests must be used.

### 2.2.1  Parallel Bars

The bars are covered by ASTM A722. They have been used extensively in prestressed concrete structures. When used in cables, they have to be coupled and their fatigue strength is relatively low.

In order to improve the fatigue strength, the anchorage is split into a dynamically acting part and a statically acting part.

The ultimate strength is normally 1,030 $N/mm^2$ (150 ksi) but may up as high as 1,470 $N/mm^2$ (213 ksi).

$E = 200,000$ $N/mm^2$ (29,000 ksi)

### 2.2.2 Parallel Seven Wire Strands

The strands have been used for a long time in prestressed concrete. Those used in cables normally have a diameter of 15 mm (0.59").

Like bars, the bundle of strands are as a rule assembled in the field. The cable may have a GUTS as high as 1,870 N/mm$^2$ (270 ksi). If the wires in the strands are galvanized, the GUTS has a lower value.

The plain seven wire strands are covered by ASTM A416 and A779.

$E = 1.8 - 1.9 \times 10^5$ N/mm$^2$ (26,000 - 28,000 ksi)

### 2.2.3 Parallel or Semi-parallel Wire Cables
(Typical Values)

Parallel bright wire cables have a GUTS of 1,670 N/mm$^2$ (242 ksi) and covered by ASTM A421.

$E = 200,000$ N/mm$^2$ (29,000 ksi)

Galvanized parallel wire cables have a GUTS of 1,570 N/mm$^2$ (228 ksi) and $E = 190,000$ N/mm$^2$ (27,500 ksi) based on Gross Area.

Long lay galvanized wire cables have GUTS of 1,570 N/mm$^2$ (228 ksi) and $E = 190,000$ N/mm$^2$ (27,500 ksi).

These three types of cables are shop assembled and brought to the site coiled on drums.

### 2.2.4 Locked Coil Strands and Ropes

The locked coil strand has several layers of round wires whereas the locked coil rope has several layers of differently shaped wires.

Only the outer layer may be galvanized or all wires may be galvanized. These cables should be prestretched to about 55% GUTS.

GUTS = 1,570 N/mm$^2$ (228 ksi)
$E = 170,000$ N/mm$^2$ (25,000 ksi)

2.2.5     Helical Galvanized Bridge Strands

The GUTS is only 670 N/mm$^2$ (97 ksi) and E = 1.6 - 1.65 x 10$^5$ N/mm$^2$ (23,000 - 24,000 ksi) depending on the strand diameter. The strand should be prestretched to about 55% of GUTS.

Its fatigue strength is low and it can only be used in cable stayed bridges where the live load is very small in comparison with the dead load.

## 2.3 CORROSION PROTECTION

Different types of corrosion protection are used for the various cable systems.

1. Parallel bars are inserted in polyethylene or steel tubes which are then filled with cement grout or some other corrosion resistant substance. The bars may also have an epoxy coating.

2. Parallel seven wire strands are treated like parallel bars.

3. Parallel or semi-parallel wire cables will normally be placed in a polyethylene tube which is filled with cement grout or another substance if the wires are not galvanized. A polyethylene sheathing is used for galvanized wires.

4. Locked coil and strands and ropes. These cables are usually left without any cover added.

5. Helical galvanized bridge strands. The cables may consist of a single strand or several parallel strands spread apart. No other protection is provided.

## 2.4 ANCHORAGES

The general requirements is that the cables fail before the anchorage, the loading being static or dynamic.

The anchorage for the bars is very simple: nuts and bearing plates.

Special anchorages are required for cables made up of parallel seven wire strands. These may be similar to those for prestressing cables or as described under b. below.

For cables made up of parallel wires, there are two basic systems:

    a.   Steel sockets filled with zinc. These anchorages have been in use for a long time. If they are improperly designed the cable tends to break at the mouth of the anchorage. This can be improved by providing a smooth rounding where the cable exists. A vibration damping device, such as a neoprene washer rigidly supporting the cable a short distance away from the anchorage also improves the cable's fatigue life.

        Having been out of favor for some years, there seems to be a return to the zinc filled socket of improved design, one reason being its low cost.

    b.   Steel sockets filled with epoxy into which small steel balls of various sizes are embedded. These balls are supposed to provide a wedging action between the wires. The latter are provided with button heads at each end which rest against a perforated bearing plate. They have a high fatigue resistance and have been widely used. The disadvantage is high cost and the fact that epoxy apparently melts at about 80 degrees Celsius (180 degrees F).

        Cable connections at tower and superstructure require careful design. It has to be assumed that water cannot be completely kept out of the cables or connections and this water must be allowed to escape. The transition between cable and anchorage should have a certain flexibility to avoid high bending stresses caused by live load and wind. At the same time, a device to prevent cable vibration from taking place at the anchorage should be provided.

Since corrosion cannot be totally prevented, visual inspection of the cable-anchorage connection is essential.

Cable-stayed bridges should be designed so that the failure of one stay or anchorage still allows the bridge to carry reduced live load.

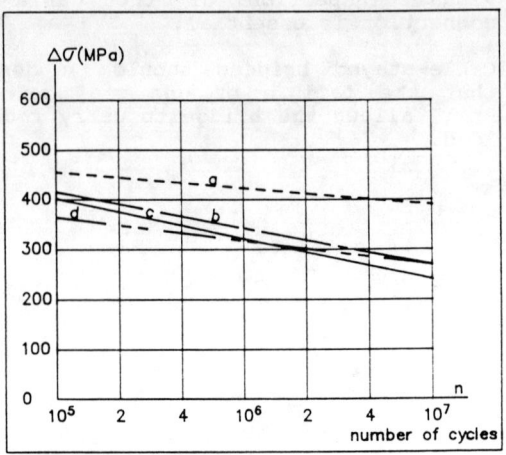

Fig. 1: Typical Wöhler's curves deduced from fatigue test at Copenhagen University Laboratory

a. single wire dia. 7 mm (.276")
b. cable with 19 wires dia. 7 mm (.276")
c. single strand dia. 15 mm (.591")
d. cable with 7 strands dia. 15 mm (.591")
    1 MPa = .145 ksi

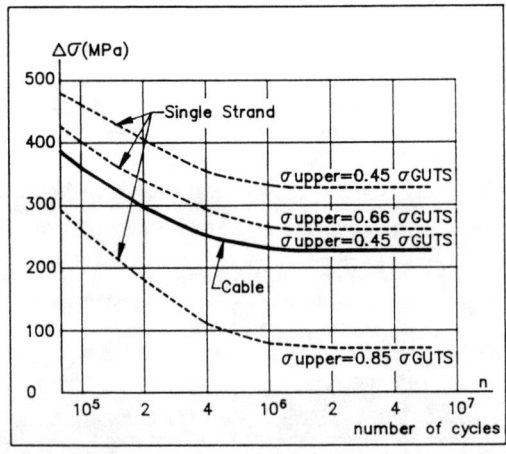

Fig. 2: Wöhler's curves for a single strand dia.
15 mm (.591") and a stay cable with Freyssinet anchorages.

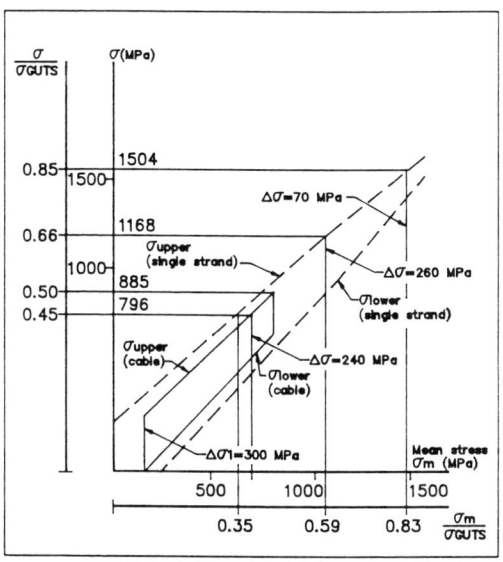

Fig. 3: Compared Smith's diagrams for a single strand dia. 15 mm (.591") and a stay cable. Number of cycles: $2.10^6$. 1 MPa = .145 ksi

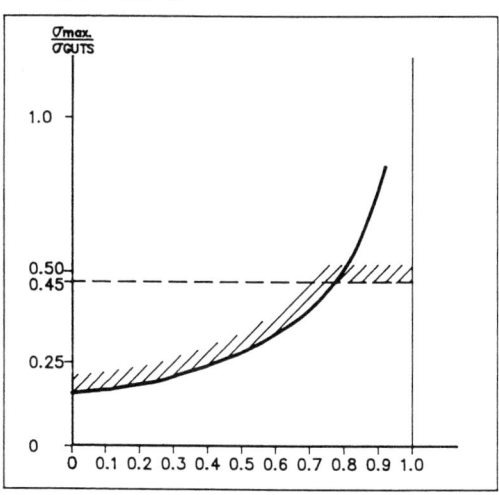

Fig. 4: Maximum stress in terms of $X = \frac{\sigma_{lower}}{\sigma_{upper}}$

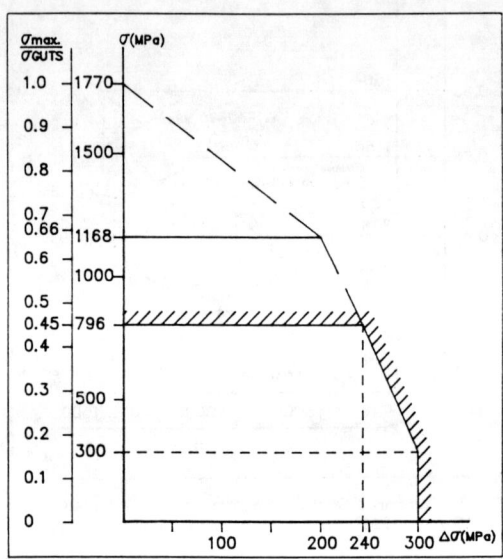

Fig. 5: Maximum stress in terms of $\Delta\sigma = \sigma_{upper} - \sigma_{lower}$
The Figures 1 through 5 have been reproduced by the permission of the Freysinnet Company.
The curves do not include a factor of safety.
1 MPa = .145 ksi

# CHAPTER 3

# STABILITY OF CABLE-STAYED BRIDGES

## 3.1 GENERAL

This Chapter deals only with stability under static loadings. For aerodynamic stabilities and stability of structure under seismic loading, see the appropriate chapters.

Both the towers and the girder of a cable-stayed bridge are compression members. But these elements also carry significant bending stresses. The P-delta effect, then is basically a nonlinear behavior which usually increases the stresses in the structural elements. It is not directly a buckling problem.

Due to the nonlinear effects, stresses and applied loads are no longer proportional to each other. The increase in stresses and deformations are usually faster than the increase of applied loads. Principle of superposition is not valid.

The nonlinear effect should be considered in both the global analysis and the local analysis.

## 3.2 GLOBAL NONLINEAR ANALYSIS

### 3.2.1 Nonlinear Behavior

The forces and deformations in a global analysis shall consider the nonlinear behavior of the structural system.

There are three major causes of nonlinearity in the global analysis:

1. Nonlinear behavior of the cable.

2. The P-delta effect of the girder and the towers.

3. Material nonlinearity.

### 3.2.2 Cable Nonlinearity

If represented by a chord member, the instantaneous stiffness of a stayed cable is [1]

$$EA_{(mod.)} = EA \cdot \left[ 1 + \frac{G^2 \cos^5\alpha \; EA}{12H^3} \right]^{-1}$$

where:  G = Total weight of cable
A = Cross-sectional Area
E = Young's Modulus
H = Horizontal component of cable force
$\alpha$ = Angle between cable chord and horizontal

When the cable force changes and its horizontal component varies from $H_1$ to $H_2$ under a specific loading, the effective cable stiffness can be calculated as

$$EA_{(mod.)} = EA \cdot \left[ 1 + \frac{G^2 \cos^5\alpha \; EA \; (H_1+H_2)}{24H_1^2 H_2^2} \right]^{-1}$$

This variation can be applied in the analysis iteratively.

### 3.2.3 The P-Delta Effect

The P-delta effect (Geometric Nonlinearity) may be considered by using the nonlinear equations for beam columns. However, it is sufficiently accurate to do the calculation iteratively using the deformed structural shape.

The proper slope of the cable shall also be adjusted in each operation due to the displacement at the cable anchorage points.

### 3.2.4 Material Nonlinearity

The material nonlinearity can usually be neglected in the analysis of cable-stayed bridges under normal loadings. However, for exceptional loadings or if the limit state of the cable-stayed bridge is to be calculated, the nonlinear relationship between moment and rotation under the influence of axial forces due to yielding of the steel and/or cracking of concrete shall be considered by applying the actual effective stiffness of the structure iteratively in the analysis.

Slackening of cables especially for bridge strands when the cable force is above the pre-stretching force shall be considered.

### 3.2.5 Base Structural System

A base structural system is the shape of the structure after all loadings are removed from the structure, including cable adjustment forces. In a rigorous nonlinear analysis all loads must be applied to the base structural system simultaneously.

### 3.2.6 Analysis

It is recommended to increase all loads including cable adjustment forces to 120% of their intended values in a nonlinear analysis. This is to assure that inaccuracy in estimating the weight of the structural elements as well as overloading will not cause excessive stresses because of the nonlinear behavior of the structure.

This procedure is to be applied to both allowable stress design and load factor design. For load factor design the value of 1.20 is used in lieu of the load factors.

## 3.3 BUCKLING LOAD

### 3.3.1 System Safety

The safety factor ($\gamma$) against elastic buckling of a cable-stayed bridge may be calculated for a specific loading to [2,4]

$$\gamma = \frac{\int EIW'' ds + \Sigma(C^2_n L/EA_{(mod)})}{\int PW'^2 ds}$$

where:  W = Mode shape ordinates
 C = Cable forces
 L = Cable length
 A = Cross-sectional Area
 E = Young's Modulus
 I = Moment of Inertia
 P = Axial force in member (compression positive)
 (') = Denotes derivative with respect to s

The above formula does not consider the nonlinear behavior of the material properties.

For the design of the towers and girder, analogous to a simple beam, the magnified design bending moment may be calculated by [2,4]

$$M_{magn} = \delta \cdot M = M \cdot \left(1 - \frac{P}{\phi P_c}\right)^{-1}$$

or with $\gamma$ available

$$\delta = \frac{\phi \gamma}{\phi \gamma - 1.0}$$

where $\delta$ is the magnification factor, P is the design axial force, Pc is the critical axial force at the referenced cross-section, $\phi$ is the strength factor and $\gamma$ is the safety factor against buckling.

### 3.3.2   Local Areas

The effect of axial compression force shall be considered in the design of local elements such as the tower walls, bridge deck, webs, and flanges. The increase in axial stresses due to the nonlinear behavior of the global action must also be included in the design of these elements.

### 3.3.3   Bridge Decks

The P-delta effect of the deck slab shall be considered in the design. Floor beams or stringers can be considered as elastic support. The axial compression generally increases the local bending moment. This effect is especially important for the design of orthotropic decks.

### 3.3.4   Towers

Thin tower walls under high compression shall be designed to resist a magnified local moment.

### 3.3.5   Cable Anchorages

Restrained cable anchorage components such as fixed anchorage can be very sensitive to nonlinear action of the cables. Change in cable forces in combination with relative rotation between the girder or tower and the cable may produce high local stresses. Maximum stresses may be calculated by means of compound influence lines.[3]

## 3.4 REFERENCES

1. Tang, M.C., "Analysis of Cable-Stayed Girder Bridges," Journal Structural Division, Proc. of the American Society of Civil Engineers, ST5, 1481-1496, May, 1970.

2. Tang, M.C., "Buckling of Cable-Stayed Girder Bridges," Journal of Structural Division, Sept. 1976, ASCE.

3. Tang, M.C., "Design of Cable-Stayed Girder Bridges," Journal of the Structural Division, ASCE, Vol. 98, No. ST8, Proc. Paper 9151, Aug. 1972, pp. 1789-1802.

4. AASHTO Standard Specifications for Highway Bridges, Edition 1989.

5. Walter Podolny, Jr., John B. Scalzi, "Construction and Design of CABLE-STAYED BRIDGES" Second Edition, Wiley-Interscience Publication.

# CHAPTER 4
## DYNAMIC WIND LOADS

### 4.1 INTRODUCTION

This chapter is concerned with the motions of cable stayed bridges that arise as a result of wind action. The aerodynamic phenomena that must be considered are vortex shedding, torsional instability or flutter, buffeting, and galloping. Intermediate and long span bridges, as well as short span pedestrian bridges are prone to wind induced motion problems and the design of such bridges must include an aerodynamic investigation. The main component of the investigation would generally be wind tunnel study using a reduced scale model.

### 4.2 REQUIREMENTS OF THE AERODYNAMIC INVESTIGATION

#### 4.2.1 Modal Analysis of the Structure

A three-dimensional modal analysis of the entire structure is recommended to determine the mode shapes and frequencies of the flexural, torsional, and lateral modes of the road deck. This information is required for correct wind tunnel modeling of the bridge. A two-dimensional modal analysis of the structure, while acceptable, may result in unnecessary conservatism of the wind tunnel experimental results. The design of the wind tunnel model also requires the mass and stiffness distribution of the towers, deck and cables.

#### 4.2.2 The Wind Environment

An investigation must be undertaken to determine the wind conditions that will prevail at the bridge site over the design life of the bridge. The extreme winds and the characteristics of the wind turbulence must be estimated. This can be done using long term meteorological records for the region and taking account of local terrain roughness characteristics. This information can be supplemented by additional wind measurements at the site and by wind tunnel studies using terrain models. Consideration must be given to the effects of long over-water fetches in suppressing turbulence at the moderate wind speeds at which vortex shedding excitation may occur.

A description of the natural wind is given in Section 4.4.

### 4.2.3 Wind Tunnel Testing

There are three approaches to the wind tunnel of bridges. These are the section model, t strip model, and the full model. These techniques are described in Section 4.5. The minimum requirement for dynamic testing would be satisfied by a sectional model test. It is recommended that full model or taut strip model testing should be preceded by a sectional model test.

Mean force measurements made using a sectional model are required for analytical estimates of buffeting response. The buffeting response may also be determined by a wind tunnel test using a full aeroelastic model.

### 4.2.4 Erection Phase Behavior

For cable stayed bridges, the tower and deck motion at various stages of erection must be considered. Either a buffeting analysis using sectional model measurements or a full model test must be done. Erection phase behavior is discussed in Section 4.7.

### 4.2.5 Components

Consideration must be given to severe motion of cables, hangers, truss members, and appendages such as luminaires signs and traffic light supports. These components may be subject to vortex shedding excitation, galloping, or torsional instability.

### 4.2.6 Structural Damping

Suitable values of structural damping must be selected. This is discussed in Section 4.8.

### 4.2.7 Probabilistic Analysis

Having determined the wind conditions under which motion occurs a probabilistic analysis over the design life of the bridge should be undertaken.

## 4.3 AERODYNAMIC EXCITATION

### 4.3.1 Vortex Shedding

Bluff sections immersed in a flow will shed vortices alternately from the upper and lower surfaces of the section. A fluctuating force develops as a result of this shedding, and if the magnitude of the fluctuating force is sufficient and occurs at a resonant frequency of the road deck, or other bridge component, motion results. Vortex shedding can excite both bending and torsional modes of bridge road decks. Ideally, a good aerodynamic cross-section geometry would eliminate motion caused by vortex shedding. This may not always be practical; recommended limits of acceleration of the deck (Reference 1) are 2% of gravitational acceleration for deck level wind speeds below 48 km/hr (30 mph) and 5% of gravitational acceleration at wind speeds above 48 km/hr (30 mph). Vortex shedding excitation occurs at wind speeds that are low enough that they are not rare events. Critical wind speeds in the range 24 to 48 km/hr (15 to 30 mph) are typical for road decks.

The vortex shedding frequency, the size of the deck, and the wind speed are related by a constant of proportionality called the Strouhal Number such that

$$S = \frac{nB}{V}$$

where n = frequency of vortex shedding, Hz
B = depth of girder plus solid part of barrier, m(ft)
V = wind speed, m/s (ft/s)

Motion commences when n = N, where N is the modal frequency of the bridge road deck in either vertical bending or torsion. Thus the critical wind speed $V_{CRIT}$ is

$$V_{crit} = \frac{NB}{S} \text{ m/s (ft/s)}$$

$$= 3.6 \frac{NB}{S} \text{ km/hr } (=.68 \frac{NB}{S} \text{ mph})$$

At wind speeds higher than $V_{crit}$ the motion amplitude increases until a maximum value is reached. Beyond this maximum the amplitudes decline until an upper limit of velocity is reached above which motion is not excited for the particular mode being considered. The amplitude of motion and the width of the velocity range over which motion occurs depend on the structural damping, the deck mass, the deck geometry, and the turbulence level in the oncoming winds.

The maximum amplitudes caused by vortex shedding are not large enough to be of concern for the safety of the bridge but are unacceptable from a user point of view. Acceptable levels of acceleration referred to earlier in this section should not be exceeded. In the case of torsional motion the vertical acceleration at the deck edge should be considered. In general, aerodynamically stable deck cross-sections have a low depth:width ratio and a degree of edge streamlining. Normally, turbulence in the natural wind reduces the amplitude of motion from that observed in smooth flow by a factor of about two or more; however wind blowing over a long fetch of water may have very low turbulence intensity because of thermal effects, and the deck response may be about the same as in smooth flow.

Determination of the response to vortex shedding requires wind tunnel testing. Wind tunnel testing approaches are discussed in Section 4.5. Sectional model tests in smooth flow are considered to give conservative estimates of the susceptibility of a road deck to vortex shedding excitation.

Components of bridges are also subject to vortex shedding excitation. This normally is not a problem with cables because of their high density. However, long slender truss members or structural member hangers such as H and I beams cause problems. Strouhal Numbers for some structural shapes are given in Table 1 of the chapter on Safety Factors, Loads, and Impact.

### 4.3.2 Torsional Instability and Flutter

Both of these instabilities begin at a critical wind speed that is normally well above those at which vortex shedding excitation occurs. The amplitude of motion reaches catastrophic values so that it is necessary to ensure that the critical speed is well in excess of the maximum design wind speed.

Torsional instability is a single degree of freedom motion whereas flutter is a coupled motion of torsion and vertical bending. Road decks are susceptible to flutter if the ratio of torsional natural frequency to bending natural frequency, $N/N_z$, is slightly higher than unity. If the ratio is higher than about 1.5 flutter would not be expected but torsional instability may still occur.

Unlike vortex shedding excitation the motion is not particularly sensitive to structural damping; increased damping levels give only small increases in critical wind speeds. However, road deck cross-section geometry is very important and modest changes can give large increases in critical wind speeds. As with vortex shedding, decks that are wide compared to their depth and have a degree of edge streamlining are desirable. It has often been found that the effect of natural wind turbulence is to increase critical speeds substantially or to eliminate the instability.

The critical wind speeds for torsional instability and flutter are proportional to the modal natural frequency in torsion. However, flutter is also dependent on the frequency ratio as stated earlier. For cable-stayed bridges, the torsional stiffness is higher with an A-frame tower, as opposed to an H-frame tower.

The intensity of natural wind turbulence depends on the nature of the upstream terrain, the rougher the terrain the higher will be the turbulence intensity and the greater the relief from these instabilities. At the high wind speeds at which these phenomena occur the thermal effects of wind blowing over a long water fetch are insignificant and the turbulence intensity will be comparable to wind blowing over flat, open terrain.

The critical wind speeds for torsional instability and flutter should be determined by wind tunnel testing. As with vortex shedding excitation, sectional model tests in smooth flow are considered to give conservative predictions of behavior.

### 4.3.3 Buffeting

The turbulence in the natural wind acts upon the bridge as fluctuations of wind speed with time. This results in a time varying wind load or buffeting. The wind speed also varies spatially. The variation along the span of the bridge is important as it results in a degree of averaging of the buffeting effect.

For the complete bridge, buffeting becomes important at high wind speeds. However, for cable-stayed bridges, serious buffeting motion may occur during erection at wind speeds that are well below the design wind speed for the completed bridge.

The magnitude of the response of the bridge depends on the properties of the turbulence, particularly the turbulence intensity, and on the shape of the road deck and the natural frequencies of the bridge.

The buffeting response of a bridge can be measured in a wind tunnel using a full aeroelastic model (Reference 2) or it can be calculated using mean aerodynamic forces measured in a wind tunnel using a rigid section model (Reference 3).

### 4.3.4 Galloping

Like torsional instability and flutter, galloping is a violent instability that occurs when a critical wind speed is exceeded (Reference 4). It excites a bending mode of vibration that is in a plane transverse to the wind direction. In general, bridge road decks are not susceptible, as only very bluff shapes are affected. For rectangular cylinders excitation will not occur if the depth in the wind direction is greater than four times the cross-wind dimension. Galloping should, therefore, be considered with pedestrian bridges.

Galloping is a more important consideration for bridge components. I-beams, H-beams, or box sections used as hangers or truss members are susceptible to galloping. Severe motion has been observed for beams with a slenderness ratio of about forty (Reference 5).

### 4.4 WIND DESCRIPTION

The magnitude of the mean wind speed depends on the averaging time used in the measurement. The mean wind speed averaged over a time T has been related to the mean hourly wind speed, $V_{MH}$, by Durst (Reference 6) using full scale wind observations.

The term fastest mile, $V_{FM}$, is widely used in the U.S.A. It is defined as a speed in miles per hour determined by the time required for one mile of air to pass a fixed point. It will be seen that the measuring time varies with speed as follows:

$$T = \frac{3600}{\text{fastest mile}} \text{ (seconds)}$$

It can be shown from the data in Reference 6 that the fastest mile and the maximum mean hourly wind can be related with sufficient accuracy by

$$V_{MH} = 0.80 \ V_{FM} \quad \text{at } z = 10 \text{ m (30 feet)}$$

The "Return Period" R is universally used as a measure of the mean interval in years between occurrences of an annual maximum wind speed. The 100 year return period annual fastest mile wind speeds are shown, as indicated earlier, in Figure 1 of the chapter on Safety Factors, Loads, and Impact.

By definition of the return period, $\frac{1}{R}$ is the probability that a particular wind speed will be

31

exceeded in any given year. Therefore, if the anticipated life span of the bridge is N years, the risk that the maximum wind speed will be exceeded in this period is given by

$$r = 1 - \left(1 - \frac{1}{R}\right)^N$$

It follows, for example, that for N = 100 years and R = 100 years we get

$$r = 0.63$$

For small values of risk, r, much larger values of R are required and for N = 100 years and r = 0.01 we get R = 9950 years. Meteorological data have been collected over a relatively short length of time and extreme value theory must be used to determine suitable values for the design wind speeds with large values of R (Reference 7).

The variation of mean wind speed, $\bar{V}(z)$, with height, z, is important and may be determined as follows:

$$\frac{\bar{V}(z)}{\bar{V}(10)} = \left(\frac{z}{10}\right)^\alpha \qquad \text{for } z \geq 10 \text{ m}$$

$$\left(\frac{\bar{V}(z)}{\bar{V}(30)} = \left(\frac{z}{30}\right)^\alpha \qquad \text{for } z \geq 30 \text{ ft}\right)$$

$$V(z) = V(10) \qquad \text{for } z \leq 10 \text{ m}$$

$$\left(V(z) = V(30) \qquad \text{for } z \leq 30 \text{ ft}\right)$$

The exponent, $\alpha$, depends on the roughness of the terrain in the upstream fetch. It can vary from 0.12 over smooth surfaces to 0.40 over very rough terrain or urban centers (Reference 8). In general a suitable value for bridge design is $\alpha = 0.20$.

The variation of gust wind speed, V(z), can be taken as

$$\frac{\overline{V(z)}}{\bar{V}(10)} = \sqrt{2} \left(\frac{z}{10}\right)^{0.1}$$

$$\left(\frac{\overline{V(z)}}{\bar{V}(30)} = \sqrt{2} \left(\frac{z}{30}\right)^{0.1}\right)$$

The mean wind speed may also depend on local topographical features such as hills or valleys. The effects may be to augment or diminish what would otherwise be selected as the design wind by as much as 50 percent. The effect on gust speed will be less and can be ignored.

The wind turbulence is described by several parameters that are statistical means averaged in the three orthogonal directions x, y, z, where x is the streamwise direction, y the lateral direction, and z vertical. Those of importance in bridge behavior are:

(i) turbulence intensity

The turbulence intensity is a measure of the fluctuating component of the wind velocity expressed as a fraction of the mean velocity. The fluctuating component, v, is defined as

$$v(t) = V(t) - \overline{V}$$

where V(t) is the instantaneous wind velocity and $\overline{V}$ is the mean wind velocity. Turbulence intensity, v', is then defined as

$$v' = \frac{(\overline{v^2})^{1/2}}{\overline{V}}$$

while v' has x, y, and z components, it is the streamwise x component that is most commonly reported. However, buffeting response depends on the magnitudes of both the x and z components.

The streamwise component of turbulence can vary from about 3% over cold water to as much as 40% over rough terrain.

(ii) turbulence scale

The turbulence scale is a measure of the mean physical size of disturbances. The scale depends on the terrain roughness and the height above ground. Typical values for the three components of scale are

33

$L_x$ = 150 m (492 ft)
$L_y$ = 30 m ( 98 ft)
$L_z$ = 0.4z

(iii) power spectral density function

The power spectral density functions, $S_x(n)$, $S_y(n)$, $S_z(n)$, are a measure of the distribution of energy in the wind as a function of the frequency. The integral of the spectrum over the entire frequency range is equal to the mean square value of the fluctuating component of velocity v, that is

$$\int_0^\infty S(n)\,dn = \overline{v^2}$$

The properties of turbulence are all dependent on height and terrain. In doing a buffeting analysis these considerations must be taken into account as shown in References 2 and 3 or when simulating the natural wind in a wind tunnel for the purpose of determining buffeting response.

The properties of atmospheric turbulence are reviewed in References 9 and 10.

## 4.5 WIND TUNNEL TESTING

There are three well-established approaches to wind tunnel testing, the sectional model, the taut strip model and the full model.

### 4.5.1 Sectional Models

This is the most frequently used method (References 11, 12, and 13). With this technique a short length of the central part of the main span is modeled. The model should be at least four times the deck width in length. The model itself is rigid, but it is suspended between the walls of the wind tunnel or between end plates, on springs that restrain torsion and vertical translation of the deck. The model to full scale size ratio is typical in the range $\lambda_L$ = 1:30 to 1:200. The deck geometry, the mass, the polar mass moment of inertia, and the frequency ratio $N_\theta/N_z$, must be modeled correctly. $N_\theta$ and $N_z$ are the natural frequencies of the bridge road deck in torsion and vertical bending. Velocity scaling, $\lambda_V$, is arbitrary but the velocity is usually scaled down by a factor of about 2 or 3. Testing is normally done in smooth flow, although testing in passive grid generated turbulence (References 14 and 15) and in actively generated turbulence has been investigated (Reference 16).

The sectional model is used to determine the susceptibility of the road deck to vortex shedding excitation, torsional instability, and flutter. Wind speeds for vortex shedding excitation and critical speeds for torsional instability and flutter can be determined and estimates of amplitude can be made from the results. The sectional model is also used to measure the mean aerodynamic force and moment coefficients acting on the deck as a function of wind angle. These data are used in estimating mean loads and in the buffeting analyses mentioned in Section 4.3.3.

The sectional model is well-suited to studying the basic aerodynamic characteristics of the deck and is a valuable tool for developing aerodynamic section improvements. It is a low cost, fast, and practical method of testing. It is limited by the omission of three-dimensional elastic effects and scaled simulation of the atmospheric turbulence. However, results have been found consistently to be conservative.

Full model tests are normally preceded by a sectional model, the combination providing the most thorough possible examination of bridge behavior.

### 4.5.2 Taut Strip Models

This method was developed to study the behavior of suspension bridges at a larger scale than possible with full models (References 17 and 18). Short segments of the deck geometry are mounted side by side, separated by small gaps, on a pair of wires. The model normally spans the wind tunnel test section. The model frequencies are controlled by the wire tension, and the frequency ratio by the spacing between the wires. Model scales are typically in the range, $\lambda_L$ = 1:100 to 1:200. It has the advantage over the sectional model that it can be tested in scaled atmospheric turbulence and that three-dimensional motion of the model is included. Some caution must be exercised as the full range of bridge motion is not simulated and for cable-stayed bridges in particular the model mode shape may not be the same as for the prototype.

This approach is a compromise between the sectional model and full model approaches. While it has its place, it has not diminished the dominance of the other two methods of testing.

### 4.5.3 The Full Model

With this approach, an aeroelastic model of the complete bridge is constructed that is dynamically similar to the prototype. It should be tested in a simulated atmospheric surface wind layer that models the conditions at the bridge site, although it is also useful to test the bridge in smooth flow when assessing vortex shedding excitation, torsional instability, and flutter. The observed motions of the bridge can be easily scaled up to give prototype response. The full model is particularly useful for determining buffeting response and critical wind speeds for torsional instability and flutter.

The geometrical scale ratio is typically in the range $\lambda_L$ = 1:100 to 1:400. For suspension bridges, gravitational forces can play a role in the dynamic behavior in wind. In order that the gravitational forces are correctly modeled, the velocity scaling, $\lambda_V$, must be equal to the square root of the geometric scale ratio. Thus

$$\lambda_V = \frac{V_{model}}{V_{prototype}} = (\lambda_L)^{1/2}$$

For cable-stayed bridges, the gravitational forces are of less importance in the dynamic behavior and the velocity scaling can be selected arbitrarily as with the sectional model.

The full model approach is the only one that includes all the relevant parameters at one time. It is the most complex, costly and time consuming but is not a good vehicle for exploratory section development. It is desirable to undertake sectional model tests in advance of the full model program.

### 4.6 SIMULATION OF THE ATMOSPHERIC BOUNDARY LAYER

There are two approaches to simulation of the earth's boundary layer. These are the spire technique, the use of floor roughness, or a combination of the two. These methods are described in References 19 and 20.

### 4.7 ERECTION PHASES

Behavior during erection phases must be considered. This is of particular importance for cable-stayed bridges erected using the balanced cantilever approach. In this case the partially erected structure will have much less stiffness than the completed structure resulting in very large stresses in the towers caused by buffeting of the road deck. This erection problem has been studied in the investigations described in References 3, 21, and 22.

The erection phase behavior can be determined using full models of the partially completed bridge or by analytical procedures. However, if results are known for the completed bridge they may be used to estimate approximately the erection phase behavior using a simplified approach. Velocities for vortex shedding and torsional instability can be related to those for the full bridge as follows:

$$\frac{V_e}{V_c} = \frac{N_e}{N_c}$$

where N is natural frequency and subscripts c and e refer to the completed bridge and the partially erected bridge respectively.

Critical wind speeds for flutter can be related by

$$\frac{V_e}{V_c} = \frac{N_{\theta e}}{N_{\theta c}} \left| \frac{1 - \left(\frac{N_{ze}^2}{N_{\theta e}}\right)}{1 - \left(\frac{N_{zc}^2}{N_{\theta c}}\right)} \right|^{1/2}$$

Response to buffeting can be related approximately by (Reference 22)

$$\frac{\sigma_{qc}}{\sigma_{qe}} = \left(\frac{V_c}{V_e}\right)^{2.75} \left(\frac{N_e}{N_c}\right)^{2.75} \left(\frac{L_e}{L_c}\right)^{0.5}$$

where $\sigma_q$ = generalized displacement in a vertical mode
L = span length

A suitable risk analysis must be undertaken for the erection phase. It may be justified to use a shorter return period than for the completed bridge, in which case there is a reduction in design wind speed and a larger reduction of the displacements because of the exponent 2.75 in the foregoing equation.

## 4.8 STRUCTURAL DAMPING

The structural damping is an important parameter in the dynamic response to wind. There have been many measurements of full scale damping of bridges (References 23, 24, and 25), although many of these have been made on short span bridges and at very low amplitude. Measured values of the critical damping ratio, $\varrho$, range from about 0.003 to well over 0.01 for both torsional and vertical bending modes and for steel and concrete bridges.

The damping value increases with amplitude and decreases with frequency. Since it is high amplitudes that are of concern, values of $\varrho = 0.01$ are commonly used in design for both bending and torsion modes.

## 4.9 REFERENCES

1. Buckland, P.G., Wardlaw, R.L, Some Aerodynamic Considerations in Bridge Design, Engineering Journal (Canada), Engineering Institute of Canada, April 1972.

2. Irwin, H.P.A.H., Wind Tunnel and Analytical Investigation of the Response of Lions' Gate Bridge to Turbulent Wind, National Research Council of Canada, NAE LTR-LA-210, June 1977.

3. Zan, S.J., Analytical Prediction of the Buffeting Response of the ALRT Fraser River Crossing to a Turbulent Wind, National Research Council of Canada NAE LTR-LA-280, January 1986.

4. Parkinson, G.V., Aeroelastic Galloping in One Degree of Freedom, Symposium No. 16, Wind Effects on Buildings and Structures, National Physical Laboratory, United Kingdom, June 1963.

5. Irwin, H.P.A.H., Cooper, K.R., Wardlaw, R.L., Application of Vibration Absorbers to Control Wind-Induced Vibration of I-Beam Truss Members of the Commodore Barry Bridge, National Research Council of Canada Report NAE LTR-LA-194, January 1976.

6. Durst, C.S., Wind Speeds Over a Short Period of Time, Meteorological Magazine, Vol. 89, No. 1056, 1960.

7.  Gomes, L. Vickery, B.J., On the Prediction of Extreme Winds from the Parent Distribution, Journal of Industrial Aerodynamics, Vol. 2, 1977, 21-36.

8.  Davenport, A.G., the Application of Statistical Concepts to the Wind Loading of Structures, Proceedings of the Institute of Civil Engineers, Vol. 19, August 1961.

9.  Panofsky, H.A., Dutton, J.A., Atmospheric Turbulence, Models and Methods for Engineering Applications, John Wiley and Sons, 1984.

10. Teunissen, H.W., Characteristics of the Mean Wind and Turbulence in the Planetary Boundary Layer, Institute for Aerospace Studies, University of Toronto, UTIAS Review No. 32, October 1970.

11. Farquharson, F.B., et al, Aerodynamic Stability of Suspension Bridges, Parts I-V, Bulletin No. 116, University of Washington, Engineering Experiment Station, Seattle, 1949-54.

12. Wardlaw, R.L, Sectional Versus Full Model Wind Tunnel Testing of Bridge Road Decks, National Research Council of Canada, DME/NAE Quarterly Bulletin No. 1978(4).

13. Scanlan, R.H., Recent Methods in the Application of Test Results to the Wind Design of Long, Suspended-Span Bridges, Report No. FHWA-RD-75-115, FHWA, U.S. Department of Transportation, Offices of R&D, Washington, D.C., October 1975.

14. Wardlaw, R.L., Tanaka, H., Savage, M.G., Wind Tunnel Investigation of the Mississippi River Bridge Steel Alternative, Quincy, Illinois, National Research Council of Canada, NAE LTR-LA-268, February 1984.

15. Davenport, A.G., King, J.P.C., The Incorporation of Dynamic Wind Loads into the Design Specifications for Long Span Bridges, ASCE Fall Convention and Structures Congress, New Orleans, Louisiana, October 1982.

16. Cermak, J.E., Bienkiewicz, B., Peterka, J.A., Active Modeling of Turbulence for Wind Tunnel Studies of Bridge Models, FHWA, U.S. Department of Transportation, Washington, D.C., Report No. FHWA-RO-82-148, February 1983.

17. Davenport, A.G., "The Use of Taut Strip Models in the Prediction of the Response of Long Span Bridges to Turbulent Wind," Proc. Symp. on Flow Induced Structural Vibrations, Paper A2, Karlsruhe, 1972.

18. Tanaka, H. and Davenport, A.G., "Response of Taut Strip Models to Turbulent Wind," Proc. ASCE Jrnl. Eng. Mech. Div. V. 108 nEM1, pp. 33-49, February 1982.

19. Cermak, J.E., Laboratory Simulation of the Atmospheric Boundary Layer, Journal of the American Institute of Aeronautics and Astronautics, Vol. 9, No. 9, September 1971, 1746-1754.

20. Irwin, H.P.A.H., Design and Use of Spires for Natural Wind Simulation, National Research Council of Canada, NAE LTR-LA-233, August, 1979.

21. Irwin, H.P.A.H., Gamble, S.L., The Action of Wind on a Cable-Stayed Bridge During Construction, Proc. 5th U.S. National Conference on Wind Engineering, Lubbock, Texas, November 1985.

22. Zan, S.J., Wardlaw, R.L, Wind Buffeting of Long Span Bridges with Reference to Erection Phase Behavior, Proc. ASCE Structures Congress '87, Bridges and Transmission Line Structures, Orlando, August 1987, pp 432-448.

23. Ito, M., Katayama, T., Nakazono, T., Some Empirical Facts on Damping of Bridges, International Association for Bridge and Structural Engineering, Reports of the working Commissions, Volume 13, Symposium, Lisbon, 1973.

24. Leonard, D.R., Eyre, R., Damping and Frequency Measurements on Eight Box Girder Bridges, Transport and Road Research Laboratory, TRRL Report 682, 1975.

25. Bampton, M.C.C., et al., Pasco-Kennewick Cable-Stayed Bridge Wind and Motion Data, U.S. Department of Transportation, Federal Highways Administration Report No. FHWA/RD-82/067, February 1983.

# CHAPTER 5

# SPECIAL CONSIDERATIONS FOR RAILROADS

## 5.0 GENERAL

In comparison with highway bridges, railroad bridges are generally characterized by a larger ratio between traffic and dead load, larger braking forces, smaller relative width of the bridge deck, and stricter requirements regarding deflections. Consequently, special care must be taken when designing cable stayed bridges for railroads.

## 5.1 Structural System

The main layout of the structural system has to be selected with due respect to the special conditions imposed by the passage of trains. Thus, the following precautions should be considered:

- o  Application of a cable system with a high degree of efficiency, e.g., a cable-stayed fan system with all stays fixed to the pylon top, or a cable-stayed harp system with intermediate supports in the side spans.

- o  A relatively small ratio between the side span and the main span length to increase the stiffness and limit the stress range in the back stays.

- o  Double support at the end of the side spans or continuity from the cable suspended side span to an adjacent beam span to limit the abrupt angular change at expansion joints.

- o  Hydraulic buffers to transmit longitudinal braking forces from the stiffening girder to the piers (without excluding slow movements due to temperature changes).

## 5.2 LOADS

### 5.2.1 Dead Loads

The dead load shall include the weight of all structural elements plus the weight of tracks, sleepers, ballast, railings, electrical installations, etc.

To account for a later increase in the weight of non-structural parts it is recommended to add a supplementary dead load of 0.5 kN/sq.m (10 lbs/ft$^2$) uniformly distributed over the entire bridge superstructure (in case this creates a more severe loading condition).

### 5.2.2 Traffic Loads

The loadings to be applied in the ultimate limit state should be as specified in the relevant specifications (AREA, UIC, etc.).

For the serviceability limit state the loadings should be chosen to represent the actual trains expected to pass over the bridge.

### 5.2.3 Wind Loads

For wind loads on railroad reference is made to existing specifications and to Chapter 1, Items 1.3 through 1.7. For dynamic wind loads see Chapter 4.

### 5.2.4 Impact

The impact formulas included in existing specifications regarding railroad loading are to be applied unless a more detailed analysis is performed. In this context it should be remembered that the existing impact formulas generally are derived by considering more traditional structures than cable suspended bridges.

## 5.3 FATIGUE

The loadings to be applied for the fatigue check should be chosen so that they represent the actual trains expected to pass over the bridge during its lifetime. Thus, the number of different train categories and their weight distribution shall be estimated.

The relatively high stress ranges induced by the heavy train loadings implies that fatigue will play a more dominant role in railroad bridges than in road bridges. It is, therefore, of special importance to choose structural elements and assemblies characterized by a high fatigue resistance.

The fatigue check should be based on a cumulative damage approach (e.g., Miners' rule).

In cases where fatigue becomes governing it will have smaller consequences for the dimensions of the structural elements if the dead load is increased and it should, therefore, be considered to use ballasted tracks on the bridge. This might also improve the dynamic behavior of the bridge.

## 5.4 DEFLECTIONS

To assure a smooth and comfortable passage of the trains on railroad bridges it is necessary to limit primarily the angular deflections in both the vertical and the horizontal plane at expansion joints (or at other points without continuity of the stiffening girder).

The allowable angular deflections will depend on the speed of the trains when passing the bridge and on the required comfort level in the trains.

As a guideline for stipulating the maximum angular slope, the following values shall be given:

For freight trains with a maximum speed of 80 km/h (50 mph):

    Vertical angular change:   maximum 0.9%
    Horizontal angular change: maximum 0.4%

For high speed passenger trains with a maximum speed of 200 km/h (125 mph):

    Vertical angular change:   maximum 0.4%
    Horizontal angular change: maximum 0.1%

In cases where these requirements cannot be fulfilled directly, a short transition span should be introduced to distribute the angular deflection to two positions.

Furthermore, it is recommended to limit the curvature in the longitudinal track profile to a radius of 10,000m (32,800 feet), and the torsional rotation at any location along the bridge to a maximum of 5%.

A limitation of the relative deflection at midspan is generally not required if a comfortable passage of the trains is demonstrated by an elaborate dynamic analysis.

## 5.5 DERAILING ACTION

The bridge deck should be designed to have sufficient strength to locally withstand wheel loads acting at a position outside the tracks.

The main structure should be designed so that at least one stay or hanger cable can be broken without increasing the maximum stresses to more than the design stress plus 10%.

## 5.6 AERODYNAMIC ACTIONS

The aerodynamic stability of a cable stayed bridge for railroad should be checked for the unloaded as well as the loaded condition. This is due to the fact that a row of railroad cars on the bridge deck might have a significant influence on the critical wind speed.

# CHAPTER 6

# SEISMIC DESIGN OF CABLE-STAYED BRIDGES

## 6.1 INTRODUCTION

For seismic design of cable-stayed bridges, the following characteristics are of importance:

1. The cable-stayed bridges consist of various components with different structural properties.

2. The fundamental natural period is generally long and the structural damping is low, as compared with girder, truss, or arch bridges having the same span length.

3. With the increase of span length, the nonlinearity and three-dimensional motion of the structure may no longer be ignored.

## 6.2 DYNAMIC CHARACTERISTICS OF CABLE-STAYED BRIDGES

### 6.2.1 Motion and Modeling of Structure

Four types of free vibration modes need to be considered in a cable-stayed bridge: vertical motion of the deck, torsional motion of the deck about its centerline, horizontal motion transverse to the span, and horizontal motion in the direction of the span if the deck has no fixed support.

The third type of motion shows the coupling with torsional motion. The predominantly torsional motion of the deck has mostly been neglected in seismic analysis. The fourth type of deck motion in the longitudinal direction is coupled with vertical motion, but the effect on the fundamental mode is not significant.

To obtain the natural frequencies and to conduct seismic response analysis, three dimensional analysis of the entire structure including substructures is advisable.

Sufficient number of masses appropriately positioned should be used in modeling of a long-span cable-stayed bridge into a lumped-mass system.

### 6.2.2 Natural Frequencies

To obtain the natural frequencies and the corresponding mode shapes, linear analysis with due considerations of the cable sag nonlinearity is generally satisfactory.

A peculiar dynamic property of cable-stayed bridges is the existence of modes with closely spaced natural frequencies.

Variations in the support conditions of the bridge deck, such as shown in Figure 1, can drastically change the fundamental natural frequency.

### 6.2.3 Structural Damping

The structural damping can only be estimated on the basis of the empirical information (7) on similar existing structures or on subjective judgement. Therefore, the collection of the full-scale measurement data is encouraged (5). Table 1 shows the values of logarithmic decrement measured on some cable-stayed bridges with steel girders.

Observations on these data are summarized as follows:

1. The damping values decrease with increasing span length. In very long spans, however, the span length is no longer influential.

2. There was no evidence of system damping (7) in multi-stay bridges from the full-scale measurements conducted.

3. The damping of the fundamental mode in torsion is generally smaller than that in bending.

4. In flexural vibrations, the damping of higher modes is lower than that of lower modes.

5. The friction at the connection of different components and at expansion joints appears to have a large influence.

Higher damping values may be used in seismic design than for wind analysis. For steel cable-stayed bridges a damping ratio of 2% for superstructures or for the entire system, 5% for piers, and 10% for foundations have been used. For prestressed concrete superstructures 3 to 5% may be assumed.

## 6.3 PROCEDURE OF SEISMIC DESIGN

### 6.3.1 General

The basic concept for the design and construction of cable-stayed bridges to resist the effects of earthquake motions is similar to that for other types of bridges. Accordingly, the AASHTO Guide Specifications for Seismic Design of Highway Bridges (1) may be referenced (this is now incorporated into the 14th Edition as mandatory). Figures 3 (a) and (b) show the design procedure flow chart provided in the above Guidelines.

Following are some commonly used methods.

### 6.3.2 Seismic Coefficient Method

The design earthquake load is the weight of the structure multiplied by the design seismic coefficient, which is the acceleration coefficient derived from elastic response spectra. This coefficient is a function of the period of the structure. Coefficient curves must be prepared according to the location of the bridge, the kind of structure, and the soil condition at the site. In general, only horizontal seismic forces are considered. But a vertical seismic force assumed as high as 2/3 of the horizontal one should be considered in the design of the connections between the superstructure and the substructure.

The resulting static earthquake forces are applied to the superstructure and tower elements as shown in Figures 5 and 6. Since several vibration modes are to be considered in the seismic design of a cable-stayed bridge, the magnitude and direction of the applied seismic coefficient may be different in different portions of the bridge, even under the same component of earthquake motion.

Stresses and displacements are then determined utilizing static structural analysis.

### 6.3.3 Response Spectrum Analysis

By combining the response spectrum and the modal analysis method (mode-superposition procedure), the maximum response of a multi-degree-of-freedom system of earthquake motion can be estimated. In this method, the maximum response can be obtained directly from the response spectrum for each individual mode of the structure. In other words, the maximum elastic-force vector in the mode i is given by (4)

$$\{F_i \max\} = [m] \{\phi_i\} \beta_i \, Sa(h_i, T_i)$$

where $[m]$ is the mass matrix of the structure, $\{\phi_i\}$ is the i-th mode shape vector, $Sa(h_i, T_i)$ is the spectral acceleration for the i-th mode, $h_i$ and $T_i$ are the damping ratio and the natural period, respectively, for the i-th mode, and

$$\beta_i = \frac{\sum_{j=1}^{n} M_j \phi_{ji}}{M_i}$$

is the modal earthquake-excitation factor or mode participation factor in which $M_i$ is the generalized mass for the i-th mode. The maximum total response is obviously less than the sum of the modal maxima because these maxima generally do not occur at the same time. Accordingly, the most prevailing practice is the use of the square root of the sum of the squares of the maximum modal responses (SRSS).

However, the SRSS method sometimes results in unrealistic seismic forces in bridges having modes with closely spaced periods. There are several methods currently available and new methods are emerging for combining these closely spaced modes. Among such new methods, the complete quadratic combination method (8) has largely replaced the SRSS in the USA and has been used in the design of some cable-stayed bridges in Japan. The CQC method has the advantage that it retains signs of the off-diagonal responses.

### 6.3.4    Time-History Response Analysis

This is done by computing the time-history response through the direct integration of the equation of motion for the structure. An actual time-history record is especially important for structures in which nonlinear response must be considered or where various types of structural interactions are involved (2).

Since records of ground motion due to very severe earthquakes at the site are usually not available, an approach to obtaining an appropriate ground-motion record has been to modify and distort an actual earthquake record so that it represents an event of different magnitude and distance. The sources of these records are:

    a. One or two ground motion records containing large accelerations and taken at other areas. It is preferable that the ground condition at the site of these earthquakes is similar to that at the bridge site.

    b. Ground-motion record taken at the bridge site, even though the intensity of the earthquake is small. The intensity of motion of these earthquakes are adjusted by means of a simple amplitude-scaling factor to have an equivalent intensity to the design acceleration.

Since the characteristics of the ground motions are random in nature and vary widely from one event to the other, it is now becoming common practice to derive artificial earthquake records which represent the desired design earthquake. One technique is to simulate with a simple mathematical model the succession of ruptures along an assumed fault line and the propagation of the vibratory waves from each successive source to the observation point.

The conventional seismic analysis does not consider the effects of multi-support seismic excitation (or non-synchronous seismic inputs) and travelling seismic waves with different speeds of propagation. These effects may become significant (2) if the span length is large.

## 6.4 DESIGN REQUIREMENTS

    1. Combination of Orthogonal Seismic Forces

        To account for the two horizontal components of earthquake motion, an analysis is required in two orthogonal directions, generally the longitudinal and transverse directions of the bridge. Forces and moments resulting from these analysis are then combined. Designating these seismic forces in a member as $F_X$ and $F_Y$ ($<F_X$), the combined design force $F_D$ can be estimated as:

$$F_D = 1.0 \ F_X + 0.3 \ F_Y$$

        according to ATC. But other combinations,

## TABLE 1
## STRUCTURAL DAMPING OBTAINED FROM FULL-SCALE MEASUREMENTS

| Name of Bridge | Main Span (m) | Mode | Bending Freq. (Hz) | Bending Log. Dec. | Torsion Freq. (Hz) | Torsion Log. Dec. |
|---|---|---|---|---|---|---|
| Yamatogawa | 355 (steel box) | 1 | 0.337 | 0.022 | 0.844 | 0.011 |
|  |  | 2 | 0.416 | 0.030 | 1.67 | 0.013 |
|  |  | 3 | 0.633 | --- |  |  |
|  |  | 4 | 0.861 | 0.029 |  |  |
| Suehiro | 240 (steel box) | 1 | 0.472 | 0.031 | 1.446 | 0.016 |
|  |  | 2 | 0.712 | 0.018 | 2.888 | 0.056 |
|  |  | 3 | 1.069 | 0.018 | 1.635 | 0.058 |
|  |  | 4 | 1.264 | 0.015 | 2.978 | 0.057 |
|  |  | 5 | 1.616 | 0.012 | 4.453 | 0.106 |
| Suigoh | 179 (steel box) | 1 | 0.454 | 0.069 | 5.630 | 0.060 |
|  |  | 2 | 0.852 | 0.037 | 6.750 | 0.090 |
|  |  | 3 | 1.256 | 0.040 |  |  |
|  |  | 4 | 2.026 | 0.064 |  |  |
|  |  | 5 | 2.558 | 0.078 |  |  |
|  |  | 6 | 3.382 | 0.063 |  |  |
|  |  | 7 | 4.589 | 0.083 |  |  |
| Alex Fraser | 465 (composite) | 1 | 0.325 | 0.028 | 0.475 | 0.019 |
| Katsushika | 220 (curved steel box) | 1 | 0.45 | 0.027 | 1.31 | 0.084 |
|  |  | 2 | 0.82 | 0.022 |  |  |
|  |  | 3 | 1.18 | 0.042 |  |  |
|  |  | 4 | 1.67 | 0.070 |  |  |
| Hitsuishi Island | 420 (truss) | 1 | 0.44 | 0.073 | 1.05 | 0.041 |
|  |  | 2 | 0.73 | 0.087 | 1.91 | 0.071 |
|  |  | 3 | 1.03 | 0.063 |  |  |
| Yokohhama | 460 (truss) | 1 | 0.34 | 0.062 | 0.88 | 0.042 |
|  |  | 2 | 0.56 | 0.181 |  |  |
|  |  | 3 | 0.80 | 0.055 |  |  |

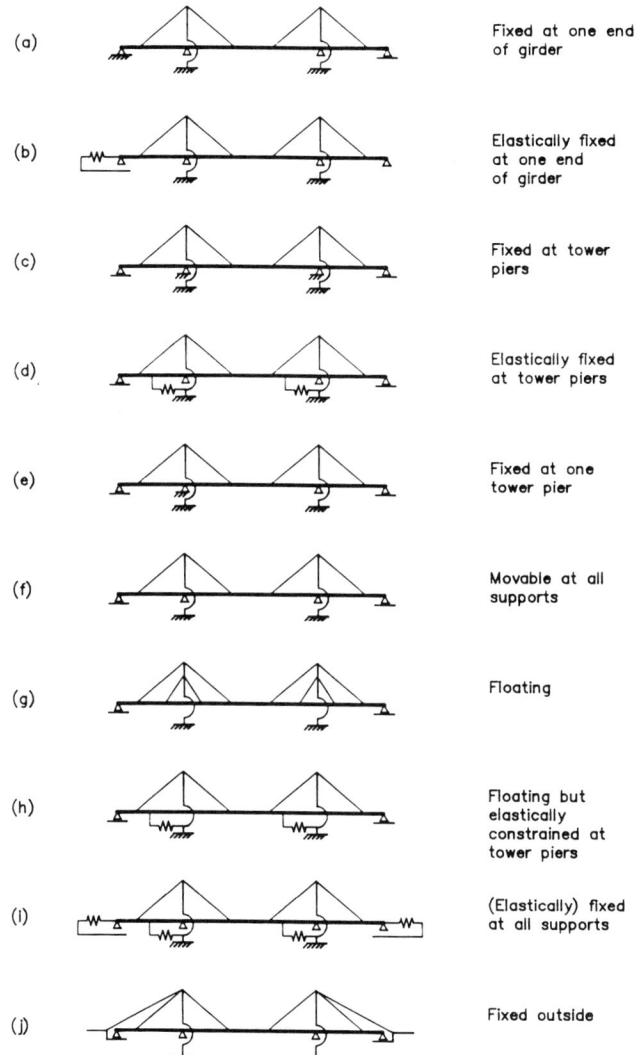

Fig. 1  Examples of different supporting conditions

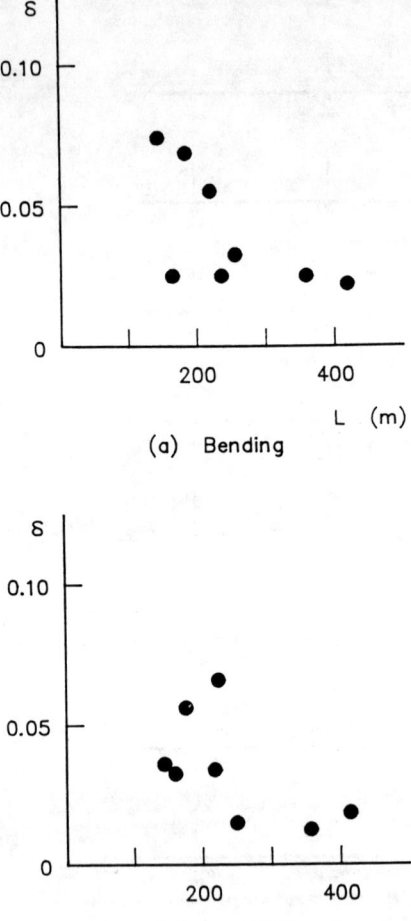

Fig. 2 Damping of Steel Cable-Stayed Bridges

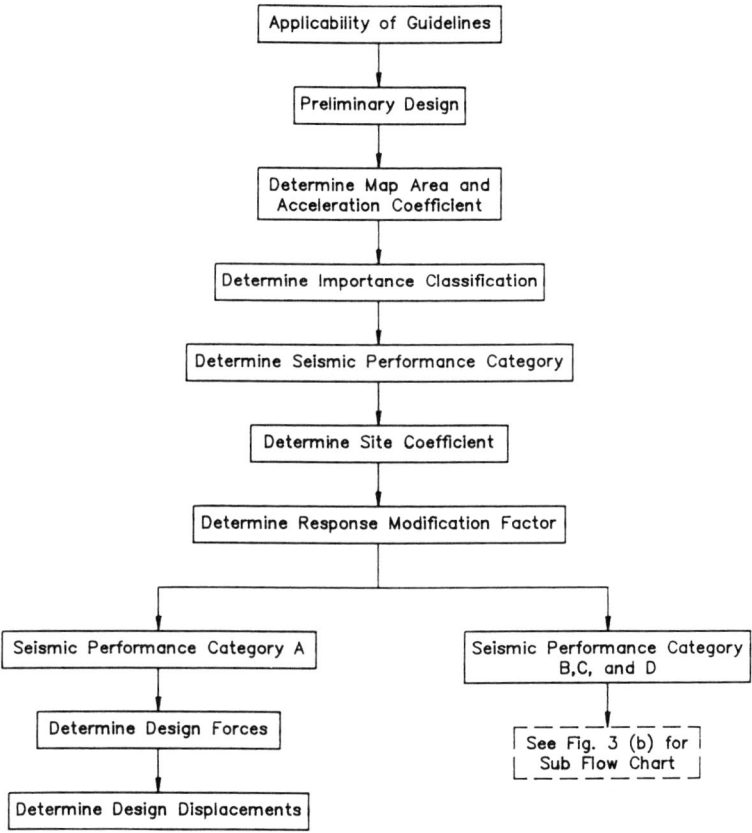

Fig. 3 (a) Design Procedure Flow Chart based on the Seismic Design Guidelines for Highway Bridges (Applied Technology Council, U.S.A.)

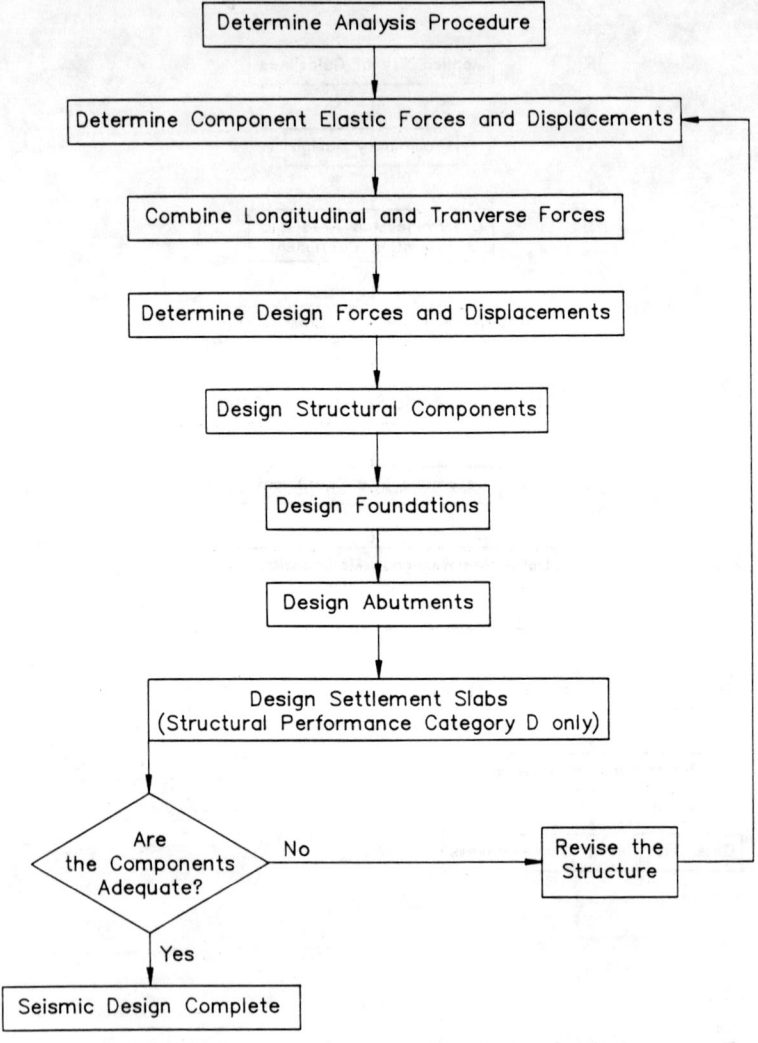

Fig. 3 (b)  Sub-Flow Chart based on the Seismic Design Guidelines for Highway Bridges (Applied Technology Council, U.S.A.)

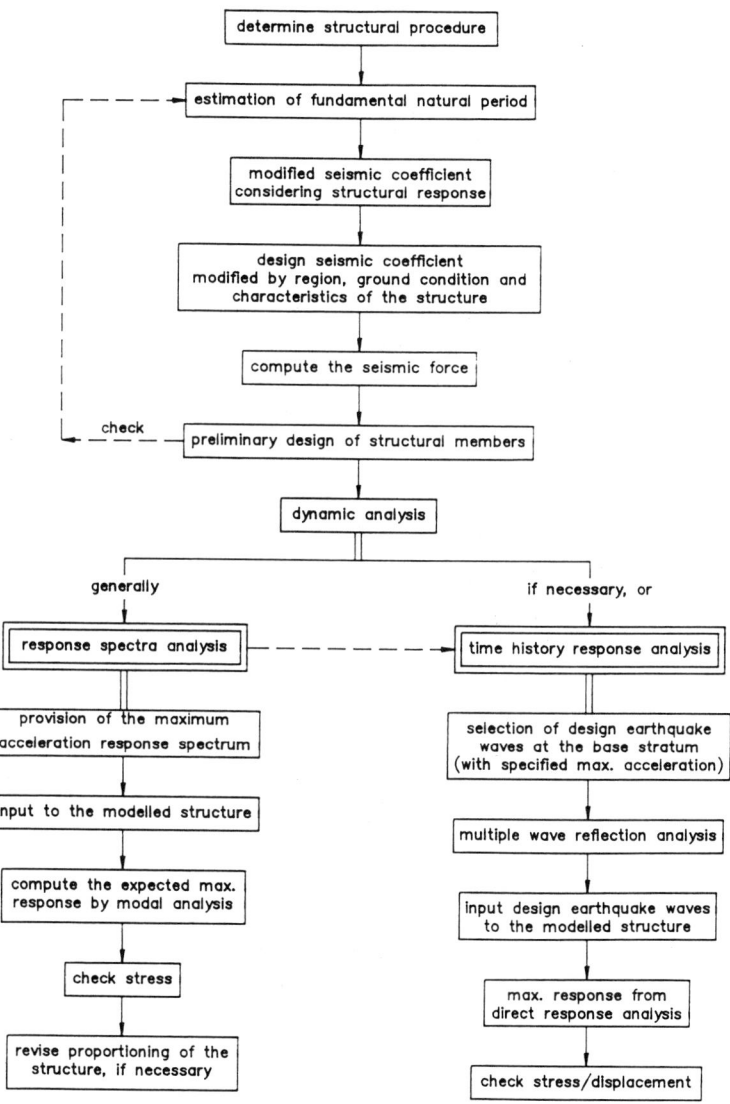

Fig. 4 Design Procedure Flow Chart for Japanese Cable—Supported Bridges.

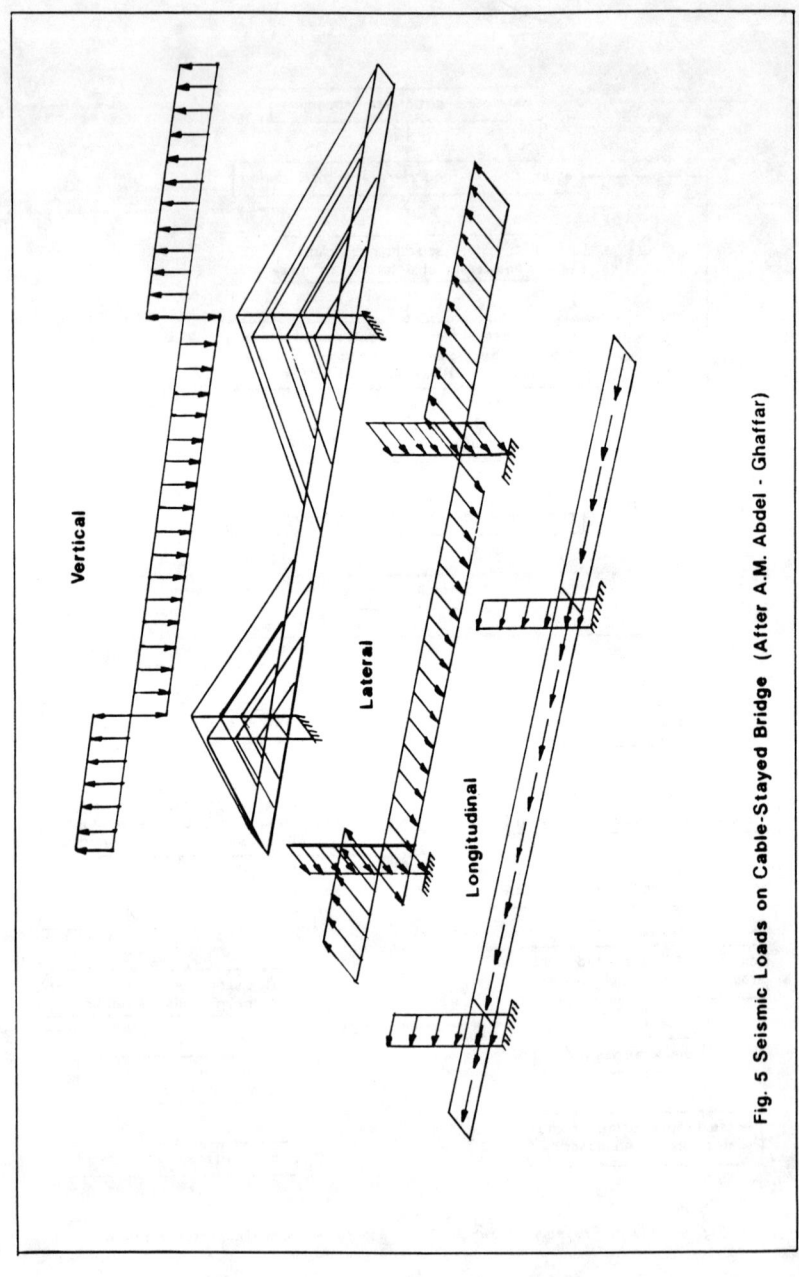

Fig. 5 Seismic Loads on Cable-Stayed Bridge (After A.M. Abdel - Ghaffar)

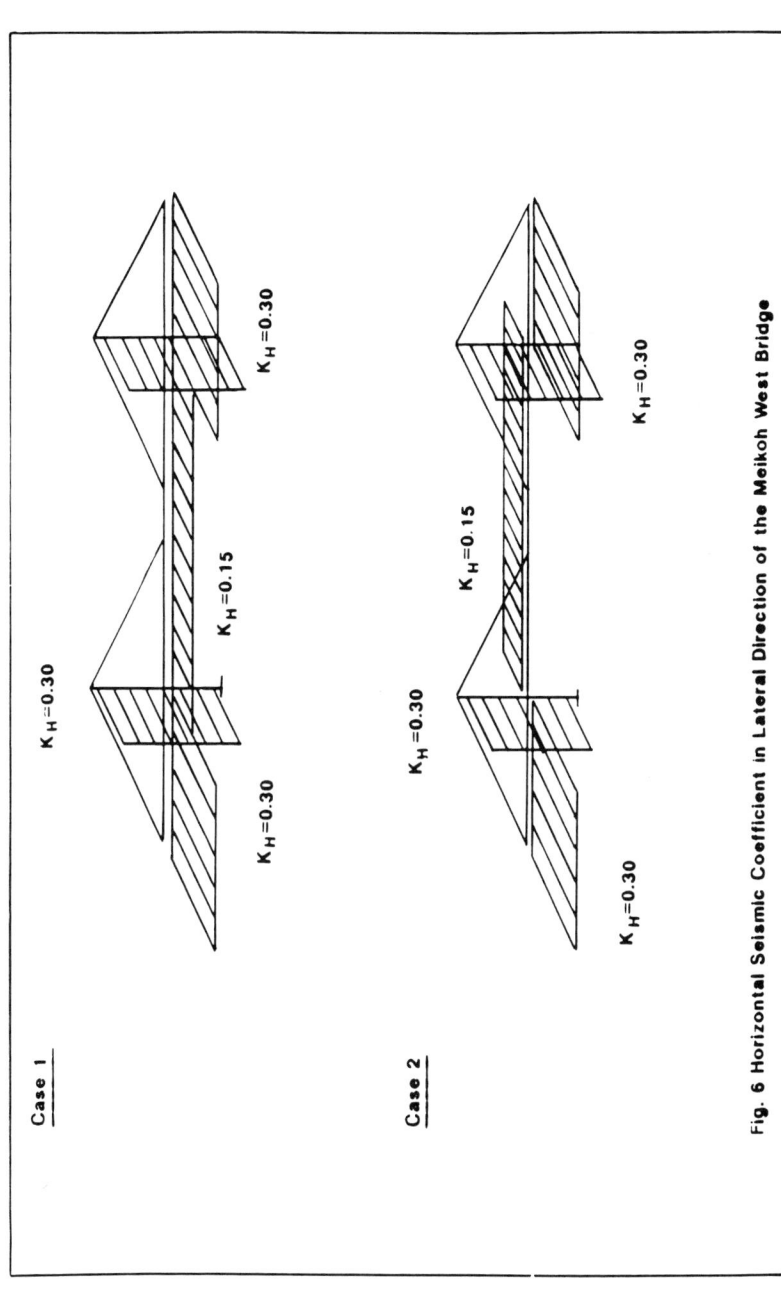

Fig. 6 Horizontal Seismic Coefficient in Lateral Direction of the Meikoh West Bridge

## Subject Index
Page number refers to first page of paper.

Aerodynamic excitation 27
Aerodynamic investigation 26
Aerodynamic stability 44
Allowable angular deflections 43
Allowable stress 1
Allowable stress design 23
Anchorages 4, 11, 12, 15, 16, 24
Atmospheric boundary layer 36, 40

Base structural system 23
Bridge decks 24
Buckling load 23
Buffeting 26, 27, 30, 33-37, 40
Buoyancy 8

Cable connections 16
Cable description 13
Cable nonlinearity 22
Cable strength 11, 12
Components 2, 3, 24, 27, 29, 31, 33, 45, 46, 49
Construction loads 8
Corrosion protection 15
Creep 13
Critical wind velocity 6

Dead loads 1, 41
Deflections 1, 9, 41, 43
Derailing 44
Design fatigue stress 11
Design procedures flow charts 55, 56, 57
Design requirements 2, 49

Earth pressure 8
Earthquakes 8, 48-50
Elastic buckling 23

Equivalent static force method 8
Erection 8, 27, 30, 36, 37
Extreme winds 4

Fastest mile 31
Fatigue 9, 11-13, 15, 16, 42, 43
Fatigue loads 9
Floating ice 7
Flutter 26, 29-31, 35-37
Free vibration modes 45
Full model 27, 34-36

Galloping 26, 27, 31
Galvanized bridge strands 11, 15
Geometric nonlinearity 22
Guaranteed ultimate tensile strength 11, 13, 14, 15
Gust wind speed 32
GUTS see Guaranteed ultimate tensile strength

Horizontal seismic coefficient 59
Horizontal wind loads 3

Impact 1, 7, 11, 13, 29, 31, 42

Linearity 9
Live loads 1, 9, 21; wind 3
Load factor design 23
Load factors 1
Loading combinations 9
Local areas 24
Locked coil strands 11, 14

Magnified design bending moment 24
Material nonlinearity 21, 22
Maximum stress 19, 20

Mean hourly wind speed  31
Modal analysis  26, 47
Mode-superposition procedure  47
Modeling  26, 27, 45
Motion  8, 26-31, 35, 40, 45, 47-49
Multimode spectral procedures  8

Natural frequencies  30, 34, 45, 46
Nonlinear analysis  21, 23
Nonlinear behavior  21, 23, 24
Normal design load  11

Orthogonal seismic forces  49

P-delta effect  21, 22, 24
Parallel bars  11, 13, 15
Parallel seven wire strands  11, 14-16
Parallel wire cables  11, 14, 15
Piers  7
Power spectral density function  34
Prestressed concrete structures  11
Probabilistic analysis  27

Railroad bridges  1, 3, 41-43
Resistance to earthquakes  50
Response spectrum analysis  47
Return period  31
Roadway bridges  1
Rotation  9, 22, 24, 43

Safety check  50
Safety factor  23, 24
Second order effects  10
Sectional models  34
Seismic coefficient method  47
Seismic design  4, 8, 45-47
Seismic loads  58
Serviceability limit state  42
Ship collision  7
Smith Diagram  12, 19
Stay forces  1

Steel sockets  16
Stream current  7
Strouhal Number  5, 6, 28
Structural damping  27-29, 38, 45, 46, 52, 54
Structural members  6
Structural stability  9
Structural system  21, 23, 41
Substructure  3, 8, 47
Superstructures  7, 46
Supporting conditions  53

Taut strip models  35, 40
Thermal forces  6
Time-history response analysis  8, 48
Torsional instability  26, 27, 29-31, 35-37
Towers  1, 8, 9, 21, 24, 26, 37
Traffic loads  42
Turbulence  30
Turbulence intensity  33
Turbulence scale  33

Ultimate limit state  42
Ultimate strength  13
Unbalanced load  8
Upward force  3

Vertical wind  3
Vortex shedding  2, 6, 26-30, 35-37

Wind description  31
Wind environment  26
Wind loads  1-3, 6, 26, 39, 42
Wind oscillations  8
Wind tunnel testing  27, 29, 30, 34
Wind turbulence  33
Wind vibration  6
Wohler Curve  12, 18

DATE DUE